Progress in Brain Research
Volume 275

From Fossils to Mind

Serial Editor

Progress in Brain Research
Volume 275

From Fossils to Mind

Edited by

Tanya Calvey
Department of Human Biology, University of Cape Town, Cape Town, South Africa

Alexandra A. de Sousa
Centre for Health and Cognition, Bath Spa University, Bath, United Kingdom

Amélie Beaudet
Department of Archaeology, University of Cambridge, Cambridge, United Kingdom; School of Geography, Archaeology and Environmental Studies, University of the Witwatersrand, Johannesburg, South Africa; Institut Català de Paleontologia Miquel Crusafont, Universitat Autònoma de Barcelona, Barcelona, Spain

ELSEVIER

Elsevier
Radarweg 29, PO Box 211, 1000 AE Amsterdam, Netherlands
The Boulevard, Langford Lane, Kidlington, Oxford, OX5 1GB, United Kingdom
50 Hampshire Street, 5th Floor, Cambridge, MA 02139, United States

First edition 2023

ISBN: 978-0-323-99107-0
ISSN: 0079-6123

For information on all Elsevier publications
visit our website at https://www.elsevier.com/books-and-journals

Publisher: Zoe Kruze
Acquisitions Editor: Mariana Kuhl
Developmental Editor: Federico Paulo Mendoza
Production Project Manager: Abdulla Sait
Cover Designer: Mark Rogers

Typeset by STRAIVE, India

Contributors

Antoine Balzeau
UMR 7194, CNRS-MNHN, Département Homme et Environnement, Musée de l'Homme, Paris, France; Department of African Zoology, Royal Museum for Central Africa, Tervuren, Belgium & Laboratory of Histology and Neuropathology, Université Libre de Bruxelles, Brussels, Belgium

Ameline Bardo
UMR 7194, CNRS-MNHN, Département Homme et Environnement, Musée de l'Homme, Paris, France; Skeletal Biology Research Centre, School of Anthropology and Conservation, University of Kent, Canterbury, United Kingdom

Amélie Beaudet
Department of Archaeology, University of Cambridge, Cambridge, United Kingdom; School of Geography, Archaeology and Environmental Studies, University of the Witwatersrand, Johannesburg, South Africa; Institut Català de Paleontologia Miquel Crusafont, Universitat Autònoma de Barcelona, Barcelona, Spain

Julien Benoit
Evolutionary Studies Institute, School of Geosciences, University of the Witwatersrand, Johannesburg, South Africa

Christine J. Charvet
Department of Anatomy, Physiology and Pharmacology, College of Veterinary Medicine, Auburn University, Auburn, AL, United States

Alexandra A. de Sousa
Centre for Health and Cognition, Bath Spa University, Bath, United Kingdom

Karina Díaz-Barba
Licenciatura en Ciencias Genómicas, UNAM, CP62210, Cuernavaca; Instituto de Ecología, UNAM, Ciudad Universitaria, CP04510, Ciudad de México, México

Kathleen N. Dollman
European Synchrotron Radiation Facility, Grenoble, France

Andréa Filippo
UMR 7194, CNRS-MNHN, Département Homme et Environnement, Musée de l'Homme, Paris, France

Danalaxshmi Shanen Ganapathee
Department of Human Evolutionary Biology, Harvard University, Cambridge, MA, United States

Pascal Godefroit
Royal Belgian Institute of Natural Sciences, Brussels, Belgium

Philipp Gunz
Department of Human Evolution, Max Planck Institute for Evolutionary Anthropology, Leipzig, Germany

Harmony Hill
Department of Archaeology, University of Cambridge, Cambridge, United Kingdom

Huseyin Kilili
Milner Centre for Evolution, Department of Biology and Biochemistry, University of Bath, Bath, United Kingdom

Veronika Kulik
Department of Anthropology and Center for the Advanced Study of Human Paleobiology, The George Washington University, Washington, DC, United States

Pascaline Lauters
Université Libre de Bruxelles; Royal Belgian Institute of Natural Sciences, Brussels, Belgium

Paul R. Manger
School of Anatomical Sciences, University of the Witwatersrand, Johannesburg, South Africa

Marta Mirazón Lahr
Department of Archaeology, University of Cambridge, Cambridge, United Kingdom

Christopher Pestana
Evolutionary Studies Institute, School of Geosciences, University of the Witwatersrand, Johannesburg, South Africa

Michael J. Proulx
Department of Psychology, University of Bath, Bath, United Kingdom

Laura D. Reyes
Department of Anthropology and Center for the Advanced Study of Human Paleobiology, The George Washington University, Washington, DC, United States

Brier A. Rigby Dames
Department of Computer Science; Department of Psychology, University of Bath, Bath, United Kingdom

Chet C. Sherwood
Department of Anthropology and Center for the Advanced Study of Human Paleobiology, The George Washington University, Washington, DC, United States

Roger M.H. Smith
Evolutionary Studies Institute, University of the Witwatersrand, Johannesburg; Karoo Palaeontology, Iziko South African Museum, Cape Town, South Africa

Orlin S. Todorov
School of Natural Sciences, Macquarie University, Sydney, NSW, Australia

Araxi O. Urrutia
Milner Centre for Evolution, Department of Biology and Biochemistry, University of Bath, Bath, United Kingdom; Licenciatura en Ciencias Genómicas, UNAM, CP62210, Cuernavaca; Instituto de Ecología, UNAM, Ciudad Universitaria, CP04510, Ciudad de México, México

Martine Vercauteren
Université Libre de Bruxelles, Brussels, Belgium

Contents

Contributors v
Preface xv

CHAPTER 1 Endocasts of ornithopod dinosaurs: Comparative anatomy 1
Pascaline Lauters, Martine Vercauteren, and Pascal Godefroit
1. Introduction 1
2. Material 3
2.1 Description of the endocranial casts 3
3. Methods 7
3.1 Telencephalon volume/endocranial volume (CRV) 8
3.2 Reptile encephalization quotient (REQ) 8
3.3 Ratio width of the cerebral hemispheres/total length of the brain. 8
4. Results 9
4.1 Olfactory peduncle 9
4.2 Pituitary gland 10
4.3 Flexures 11
4.4 Floccular fossa 11
4.5 Reptile encephalization quotient (REQ) 11
4.6 Telencephalon volume/endocranial volume (CRV) 11
4.7 Ratio width of the cerebral hemispheres/total length of the brain 12
5. Discussion 12
5.1 Optic lobes 12
5.2 Olfactory tracts 12
5.3 Flexures 13
5.4 Cerebral hemispheres and encephalization quotient 13
5.5 CRV 15
5.6 Behavior 15
Acknowledgments 17
References 17

CHAPTER 2 At the root of the mammalian mind: The sensory organs, brain and behavior of pre-mammalian synapsids 25
Julien Benoit, Kathleen N. Dollman, Roger M.H. Smith, and Paul R. Manger
1. Introduction 26

2. Sensory organs in mammalian ancestors: what could they smell, see, hear and feel? ... 28
2.1 Low-resolution olfaction? ... 29
2.2 Poor vision? ... 31
2.3 Insensitive hearing? ... 34
2.4 Coarse tactile sensitivity? ... 38
2.5 Unrefined motor coordination? ... 42
2.6 Summary of sensory organs ... 43
3. Behavioral complexity ... 44
3.1 Small brain? ... 44
3.2 Simple behavioral repertoire? ... 49
3.3 Discussion on brain and behavior ... 53
4. Conclusion: the evolution of the synapsid brain, behavior, and sensory organs, brain and behavior ... 55
Acknowledgments ... 57
References ... 58

CHAPTER 3 Insights into brain evolution through the genotype-phenotype connection ... 73
Danalaxshmi Shanen Ganapathee and Philipp Gunz
1. Why is the genotype-phenotype connection important? ... 73
2. What has our field done thus far to elucidate the genotype-phenotype connection? ... 75
2.1 Evolutionary perspectives on the human brain phenotype ... 75
2.2 Neandertal genomes offer clues about the evolution of brain shape and behavior ... 79
2.3 Mice, ferrets, and brains in a petri dish ... 80
2.4 Outstanding gaps in our understanding of human-specific brain and behavior evolution ... 83
3. What can we do as researchers into the genotype-phenotype connection to advance the unsolved riddles of human evolution? ... 84
Acknowledgments ... 87
References ... 87

CHAPTER 4 Coevolution of language and tools in the human brain: An ALE meta-analysis of neural activation during syntactic processing and tool use ... 93
Veronika Kulik, Laura D. Reyes, and Chet C. Sherwood
1. Introduction ... 94
2. Methods ... 96
3. Results ... 102
4. Discussion ... 105

4.1 Syntax activations ... 106
4.2 Tool use activations ... 107
4.3 Shared activations—Inferior frontal gyrus ... 107
4.4 Limitations and considerations ... 108
4.5 Conclusions ... 108
References ... 109

CHAPTER 5 Brain evolution and language: A comparative 3D analysis of Wernicke's area in extant and fossil hominids ... 117
Harmony Hill, Marta Mirazón Lahr, and Amélie Beaudet

1. Introduction ... 118
1.1 Broca's area in extant humans ... 118
1.2 Wernicke's area in extant humans ... 120
1.3 Homologous regions in nonhuman primates ... 120
1.4 Introducing endocasts ... 121
2. Materials and methods ... 123
3. Results ... 124
3.1 Extant chimpanzee brains ... 124
3.2 Extant human brains ... 128
3.3 Fossil hominin endocasts ... 132
4. Discussion ... 134
4.1 Variation within the extant sample ... 134
4.2 Interpreting the hominin fossil record ... 136
4.3 Limitations and conclusive remarks ... 137
Acknowledgments ... 139
References ... 139

CHAPTER 6 Lateralized behaviors in living humans: Application in the context of hominin brain evolution ... 143
Ameline Bardo, Andréa Filippo, and Antoine Balzeau

1. Introduction ... 144
2. Functional asymmetry of the human brain and lateralized behaviors ... 145
2.1 Language ... 146
2.2 Human handedness ... 147
2.3 Visuospatial skills and attention ... 149
3. How to infer functional asymmetry of fossils hominins? ... 150
3.1 Paleoneurology and endocasts ... 151
3.2 Hand preference ... 152
3.3 Directional asymmetries in arm bones ... 153

4. Prospective from anatomy to behavior: The paleobrain project...154
4.1 Complexity of human laterality...155
4.2 Endocasts: From living humans to fossil hominins...156
5. Conclusion...157
Acknowledgments...157
References...157

CHAPTER 7 Evolutionary and genomic perspectives of brain aging and neurodegenerative diseases...165
Brier A. Rigby Dames, Huseyin Kilili, Christine J. Charvet, Karina Díaz-Barba, Michael J. Proulx, Alexandra A. de Sousa, and Araxi O. Urrutia
1. Introduction...166
2. Neurodegenerative diseases and brain aging at the macroscale...168
2.1 Prevalence and classification of Alzheimer's disease and Lewy body disorders...168
2.2 Structural brain changes with neurodegeneration and typical aging...168
3. Neurobiology and genetics of aging and neurodegenerative diseases...173
3.1 Neurobiology of aging and neurodegenerative diseases...173
3.2 Genetics of aging and neurodegenerative diseases...174
4. Comparative neuropathology between human and nonhuman primates...178
4.1 Neuropathology of Alzheimer's disease...178
4.2 Neuropathology in Parkinson's disease and other Lewy body disorders...180
5. Evolutionary hypotheses of aging and neurodegenerative diseases...182
5.1 Evolutionary theories of aging...182
5.2 Species' maximum lifespan potential is tightly coupled to the evolution of brain size...183
5.3 Aging in the hominin fossil record...187
5.4 Evolutionary perspective of neurodegenerative diseases...189
6. Conclusion...192
Supplementary material...193
Acknowledgments...193
References...193